Grades 5-8

Focus On Middle School

Laboratory Workbook

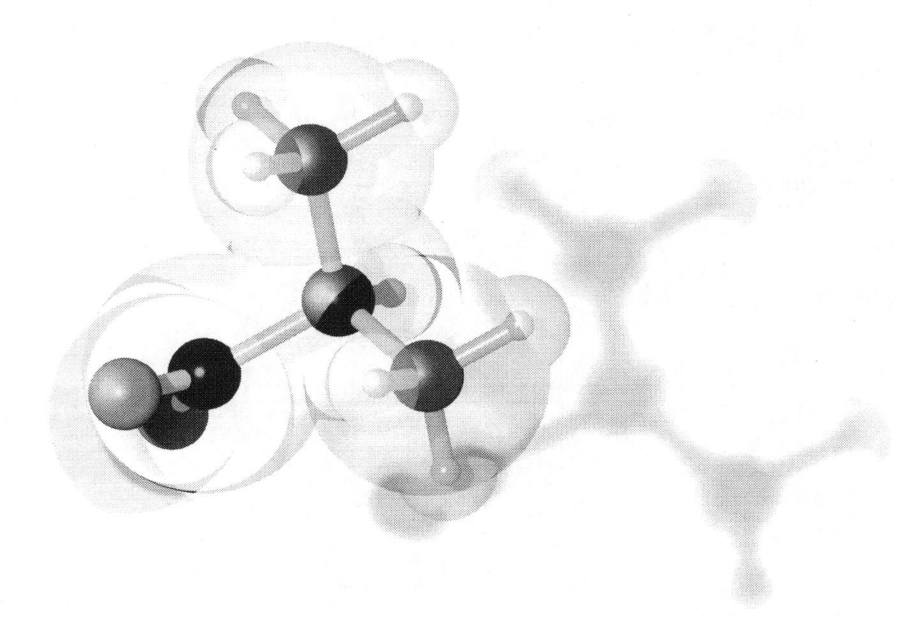

Rebecca W. Keller, PhD

Cover design: David Keller
Opening page: David Keller, Rebecca W. Keller, PhD
Illustrations: Rebecca W. Keller, PhD

Focus On Middle School Chemistry Laboratory Workbook
ISBN: 978-1-936114-60-3

Published by Gravitas Publications, Inc.
www.gravitaspublications.com

Printed in the United States

Keeping a Laboratory Notebook

A laboratory notebook is essential for the experimental scientist. In this type of notebook, the results of all the experiments are kept together along with comments and any additional information that is gathered. For this curriculum, you should use this workbook as your laboratory notebook and record your experimental observations and conclusions directly on its pages, just as a real scientist would.

The experimental section for each chapter is pre-written. The exact format of a notebook may vary among scientists, but all experiments written in a laboratory notebook have certain essential parts. For each experiment, a descriptive but short *Title* is written at the top of the page along with the *Date* the experiment is performed. Below the title, an *Objective* and a *Hypothesis* are written. The objective is a short statement that tells something about why you are doing the experiment, and the hypothesis is the predicted outcome. Next, a *Materials List* is written. The materials should be gathered before the experiment is started.

Following the *Materials List* is the *Experiment*. The sequence of steps and all the details for performing the experiment are written beforehand. Any changes made during the experiment should be written down. Include all information that might be of some importance. For example, if you are to measure 237 ml (1 cup) of water for an experiment, but you actually measured 296 ml (1 1/4 cup), this should be recorded. It is hard sometimes to predict the way in which even small variations in an experiment will affect the outcome, and it is easier to track a problem if all of the information is recorded.

The next section is the *Results* section. Here you will record your experimental observations. It is extremely important that you be honest about what is observed. For example, if the experimental instructions say that a solution will turn yellow, but your solution turned blue, you must record blue. You may have done the experiment incorrectly, or you might have discovered a new and interesting result, but either way, it is very important that your observations be honestly recorded.

Finally, the *Conclusions* should be written. Here you will explain what the observations may mean. You should try to write only *valid* conclusions. It is important to learn to think about what the data actually show and also what cannot be concluded from the experiment.

Laboratory Safety

Most of these experiments use household items. However, some items, such as iodine, are extremely poisonous. Extra care should be taken while working with all chemicals in this series of experiments. The following are some general laboratory precautions that should be applied to the home laboratory:

Never put things in your mouth without explicit instructions to do so. This means that food items should not be eaten unless tasting or eating is part of the experiment.

Use safety glasses while working with glass objects or strong chemicals such as bleach.

Wash hands before and after handling chemicals.

Use adult supervision while working with iodine and while conducting any step requiring a stove.

Contents

Experiment 1: What Is It Made Of? Date: ____________

Objective

To become familiar with the periodic table of elements and investigate the composition of some common items

Materials

pen or pencil
food labels
periodic table of elements
resources (books or online) such as:
  dictionary
  encyclopedia
computer with internet access (optional)

Experiment

❶ Using the periodic table of elements, answer the following questions:

A. How many protons does aluminum have? ____________________

How many electrons? ____________________

B. What is the symbol for carbon? ____________________

C. List all the elements that have chemical properties similar to helium.

____

____

____

D. What is the atomic weight of nitrogen? ____________________

How many neutrons does nitrogen have? ____________________

❷ In the table on the next page, fill in the following information.

- **ITEM**

  Think of several different items and write them in the column labeled **ITEM**. These can be any item, like "tires" or "cereal." Try to be specific. For example, instead of writing just "cereal," write "corn cereal" or "sweet, colored cereal."

- **COMPOSITION**

  In an encyclopedia, on the food label, or online, look up the composition of the items you have selected, and write this information in the column labeled **COMPOSITION**. Try to be as specific as possible when identifying the composition. For example, if your cereal contains vitamin C, write "sodium ascorbate" if that name is also listed. Try to identify any elements that are in the compounds you have listed. For example, vitamin C contains the element "sodium."

- **SOURCE**

  Write the source next to the composition. "Source" means where you got your information; for example, "food label" or "encyclopedia," or if you got the information online, list the name of the website.

ITEM	COMPOSITION	SOURCE
1.		
2.		
3.		
4.		
5.		
6.		
7.		
8.		

Results

Briefly describe what you discovered about the composition of the various items.

For example:

Kellogg's Sugar Smacks cereal contains vitamin C, which is called sodium ascorbate.

Conclusions

State your conclusions based on the information you collected.

For example:

Many cereals contain sodium in the form of salt and vitamin C.

Review

Define the following terms:

chemistry ______________________________

____

matter ______________________________

____

atoms (atomos) ______________________________

____

proton ______________________________

____

neutron ______________________________

____

electron ______________________________

____

nucleus ______________________________

____

electron cloud ______________________________

____

element ______________________________

____

atomic weight ______________________________

____

Experiment 2: Making Marshmallow Molecules Date:_____

Objective To make models of molecules from marshmallows and toothpicks to show how atoms fit together

Materials

small, colored marshmallows
large marshmallows
toothpicks

Experiment

1. Take several marshmallows of both sizes and several toothpicks.
2. Make shapes from the marshmallows and toothpicks. First, form any number of links between marshmallows (i.e., put any number of toothpicks into each marshmallow). Draw the shapes below, noting the number of toothpicks in each marshmallow.

❸ Using new marshmallows, assign an "atom" to each of the marshmallows. The large marshmallows should be C, N, and O, and the small marshmallows should be H and Cl. Use the following "rules" for the number of toothpicks that can go into a marshmallow.

carbon – 4 toothpicks all pointing away from each other

nitrogen – 3 toothpicks pointing downward

oxygen – 2 toothpicks pointing downward

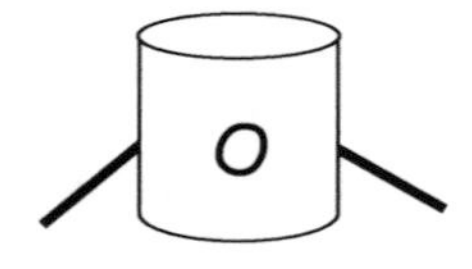

hydrogen and chlorine – 1 toothpick pointing in any direction

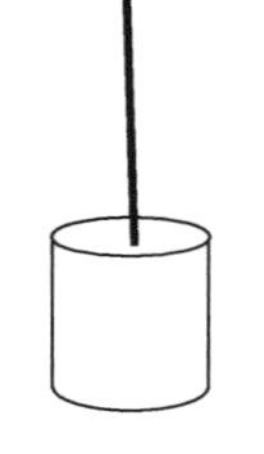

Cl or H

❹ Next, try to make the following molecules from your marshmallow atoms.

To make each molecule, follow the rules in Step 3, then draw the shape of each molecule you make on the following chart.

H_2O: This is one oxygen and two hydrogens.

NH_3: This is one nitrogen and three hydrogens.

CH_4: This is one carbon and four hydrogens.

CH_3OH: This is one carbon with three hydrogens and one oxygen attached. The oxygen has one hydrogen attached to it.

H_2O	NH_3
CH_4	CH_3OH

5. Now, following the "rules" outlined in Step 3 for the marshmallow molecules, make other "molecules." Make as many different shapes as you can without breaking the "rules." Draw your shapes below.

Conclusions

Review

Define the following terms:

molecule ______________________________

____

bond ______________________________

____

shared electron bond ______________________________

____

unshared electron bond ______________________________

____

sodium chloride ______________________________

____

Answer the following questions:

- How many bonds does hydrogen typically form? ______________
- How many bonds does carbon typically form? ______________
- How many bonds does nitrogen typically form? ______________
- How many bonds does oxygen typically form? ______________

Draw the shape of a water molecule.

Experiment 3: Identifying Chemical Reactions

Date: _____

Objective

In this experiment we will try to identify a chemical reaction by observing the changes that occur when two solutions are added together.

Hypothesis

A chemical reaction can be identified by observing changes that occur in the course of the reaction.

Materials

baking soda
lemon juice
balsamic vinegar
salt and water: 15–30 ml salt dissolved in 120 ml water (1–2 tbsp. salt dissolved in 1/2 cup of water)
egg whites
milk
several small jars
measuring cups and spoons
eye dropper

Experiment

❶ Look at the chart in the *Results* section. Write down all of the substances (baking soda, lemon juice, balsamic vinegar, salt water, egg whites, and milk) horizontally with one item above each column.

❷ Now write the same list of items vertically down the left side of the grid, next to each row.

❸ There should be an item assigned to each column and to each row.

❹ In the white boxes of the chart on the next page, record what you observe when the item listed in the column is mixed with the item in the corresponding row.

5. Look especially for changes that indicate a chemical reaction has taken place. For example, look for bubbles, color change, or a precipitate.
6. Ask your teacher for unknown solutions. When you mix them, try to determine whether a chemical reaction has taken place. Try to identify what the unknown solutions are.

Results

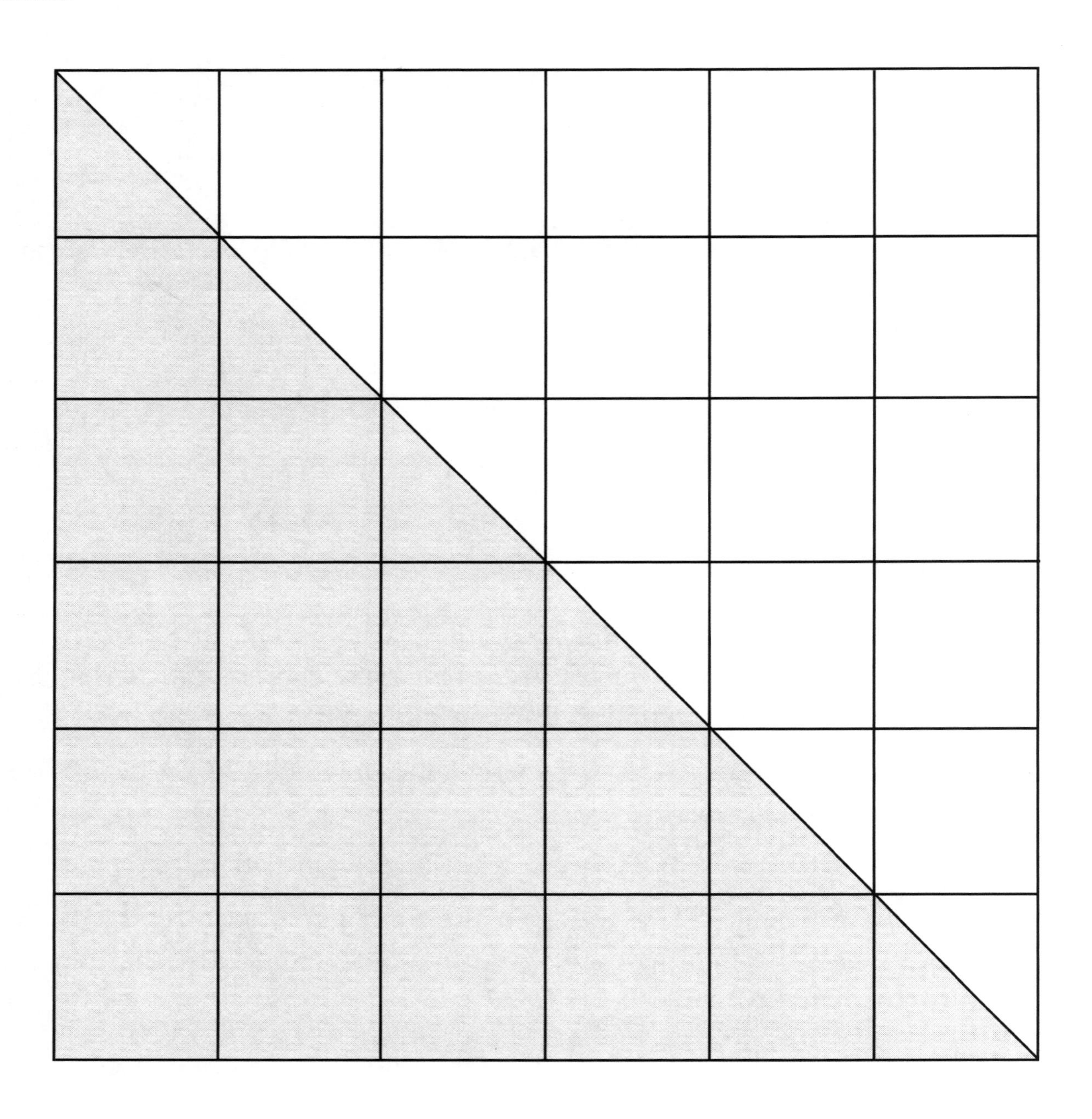

Results for Unknown Solutions

Description of each unknown solution:

❶ ______________________________

____

❷ ______________________________

____

Results when the two are mixed:

____

____

____

____

What might they be?

❶ ______________________________

____

❷ ______________________________

____

Conclusions

Review

What are the four types of chemical reactions?

____

____

____

____

Define the following terms:

chemical reaction ____

____

combination reaction ____

____

decomposition reaction ____

____

displacement reaction ____

____

exchange reaction ____

____

spontaneous ____

____

List four signs that may be observed that indicate a chemical reaction has occurred.

____

____

____

____

Experiment 4: Making an Acid-Base Indicator Date:____

Objective

____

____

Hypothesis

____

____

Materials

one head of red cabbage
distilled water
various solutions, such as:
- ammonia
- vinegar
- soda pop
- milk
- mineral water

large saucepan
knife
several small jars
white coffee filters
eye dropper
measuring cup
measuring spoons
marking pen
scissors
ruler

Experiment

❶ Take the whole head of red cabbage and divide it into several pieces.

❷ Place about 700 ml (3 cups) of distilled water in a large saucepan and bring the water to a boil. Place the cabbage in the boiling water and boil for several minutes.

❸ Remove the cabbage and let the water cool. The water should be a deep purple color.

❹ Take 240 ml (1 cup) of the cabbage water to use in this experiment, and REFRIGERATE the rest for the next experiment.

❺ Cut the coffee filters into small strips about 2 cm (3/4 inch) wide and 4 cm (1.5 inches) long. Make at least 20.

❻ Using the eye dropper, put several drops of the cabbage water onto each of the filter papers and allow them to dry. They should be slightly pink and uniform in color. If the papers are too light, more solution can be dropped onto them, and they can be dried again. These are your acid-base indicator (pH) papers.

❼ Label one of the jars **Control Acid**, and place 15 ml (1 tbsp.) of vinegar in the jar. Add 75 ml (5 tbsp.) of distilled water. This is your *known acid.*

Label another jar **Control Base** and add 15 ml (1 tbsp.) of ammonia to the jar. Add 75 ml (5 tbsp.) of distilled water. This is your *known base.*

Put 15 ml (1 tbsp.) of each of the other solutions you have collected into separate jars, and add 30–75 ml (2–5 tbsp.) of distilled water to each jar.

❽ Carefully dip the pH paper into the **Control Acid**. Look immediately at the pH paper for a color change and record your results in the chart on the next page. Then tape the pH paper in the *pH Paper Sample* column next to the section labeled *Control Acid.*

❾ Carefully dip a new piece of pH paper into the **Control Base**. Look immediately at the pH paper for a color change, and record your results in the chart on the next page. Tape the pH paper in the space next to the **Control Base** section.

❿ Now take new pieces of pH paper, and dip them into the other solutions you have made. Record your results. Tape the papers into the chart.

Results

pH Paper Sample	Name of Solution	Color of pH Paper	Acid/Base?
	Control Acid:		
	Control Base:		

Conclusions

Review

Define the following terms:

electrode ______________________________

____

pH meter ______________________________

____

litmus paper ______________________________

____

acid-base indicator ______________________________

____

acid-base reaction ______________________________

____

controls ______________________________

____

Answer the following questions.

- What is the pH of a neutral solution? ______________
- What is the pH of an acidic solution? ______________
- What is the pH of a basic solution? ______________
- Is vinegar an acid or a base? ______________
- Is baking soda an acid or a base? ______________
- What is the chemical name for vinegar? ______________
- What is the chemical name for baking soda? ______________

Experiment 5: Vinegar and Ammonia in the Balance: An Introduction to Titrations

Date: ______

Objective

____

____

Hypothesis

____

____

Materials

red cabbage indicator (from Experiment 4)
household ammonia
vinegar
large glass jar
measuring spoons
measuring cup

Experiment

1. Measure 60 ml (1/4 cup) of vinegar, and put it in the glass jar.
2. Add enough of the red cabbage indicator to get a deep red color.
3. Carefully add 5 ml (1 tsp) of ammonia to the vinegar solution. Swirl gently, and record the color of the solution in the chart on the following page.
4. Add another 5 ml (1 tsp) of ammonia to the vinegar solution, and record the color of the solution.
5. Keep adding ammonia to the vinegar solution 5 ml (1 tsp) at a time, and record the color of the solution each time.
6. When the color has changed from red to green, stop adding ammonia.

Results

Number of Ml (Tsp) of Ammonia	Color of Solution
5 ml (1 tsp)	red
10 ml (2 tsp)	red

Number of Ml (Tsp) of Ammonia [continued]	Color of Solution [continued]

7. Plot the data from your chart using the graph on the next page. The horizontal axis is labeled **Amount of Ammonia**, and the vertical axis is labeled **Color of Solution**.
8. For every 5 ml (1 tsp) of ammonia added, mark the graph with a round dot corresponding to the color of the solution.
9. When all of the data have been plotted, connect the dots.

Graphing Your Data

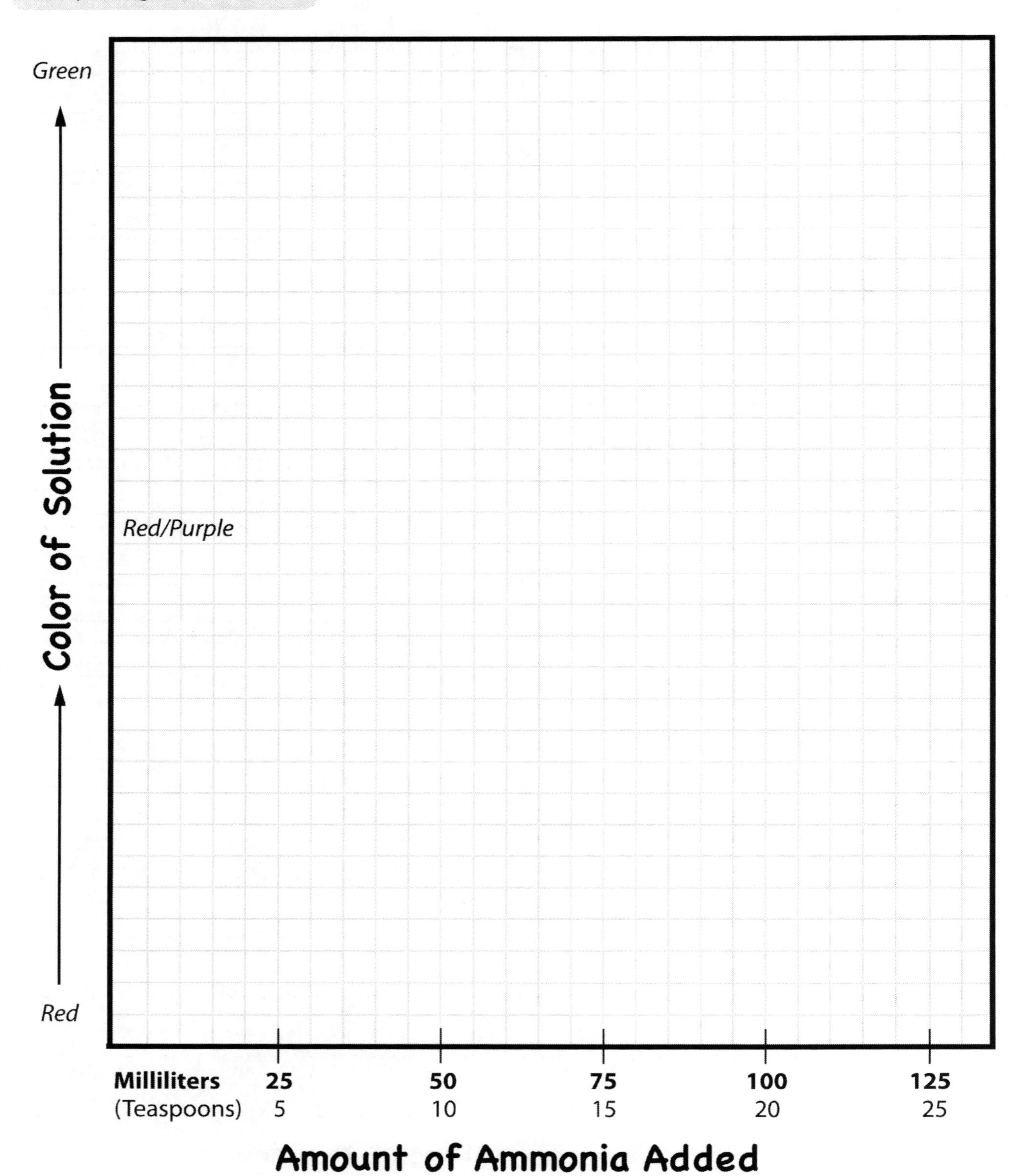

Conclusions

Review

Define the following terms:

neutralization reaction ______________________________

____

concentration ______________________________

____

concentrated ______________________________

____

dilute ______________________________

____

glacial acetic acid ______________________________

____

acid indigestion ______________________________

____

titration ______________________________

____

axes (axis) ______________________________

____

Experiment 6: Mix It Up!

Date: ________

Objective

____

____

Hypothesis

____

____

Materials

water
ammonia
vegetable oil
rubbing alcohol
melted butter
vinegar
small jars (7 or more)
food coloring
dish soap
eye dropper
measuring cup and
measuring spoons
marking pen

Experiment

Part I: See What Mixes

❶ The grid in the *Results* section is labeled along the top and one side with names of the following liquids: water, ammonia, vegetable oil, rubbing alcohol, melted butter, and vinegar.

❷ Take 6 small jars and put 60 milliliters (1/4 cup) of each liquid in its own jar, labeling each jar.

❸ Add a drop of food coloring to each jar.

❹ Using a clean jar each time, mix 15 ml (1 tablespoon) of a liquid that has not been colored with 15 ml (1 tablespoon) of a colored liquid. In the white boxes on the next page, record whether or not the two liquids mix.

Results

	Water	Ammonia	Vegetable Oil	Rubbing Alcohol	Melted Butter	Vinegar
Water						
Ammonia						
Vegetable Oil						
Rubbing Alcohol						
Melted Butter						
Vinegar						

Part II: Soap, Oil, and Water

1. Put 60 ml (1/4 cup) of water into a small glass jar. Add one drop of food coloring.
2. Add 15 ml (1 tablespoon) of vegetable oil to the water.
3. Mix the water and oil. Record your results below.
4. Add 15 ml (1 tablespoon) of liquid dish soap to the oil/water mixture.
5. Mix thoroughly. Record your results.
6. Add another 15 ml (1 tablespoon) of liquid dish soap to the mixture, and mix thoroughly.
7. Record your results.

Results

- Oil + water ______________________________

  ____

  ____

- Oil + water + 15 ml (1T) soap ______________________________

  ____

  ____

- Oil + water + 30 ml (2T) soap ______________________________

  ____

  ____

Conclusions

Review

Define the following terms:

mixture ______________________________

____

homogeneous ______________________________

____

heterogeneous ______________________________

____

dissolve ______________________________

____

Answer the following questions:

- What does the phrase "like dissolves like" mean? ______________________________

  ____

  ____

- Name two molecules with charged ends. ______________________________

  ____

- Name two molecules without charged ends. ______________________________

  ____

- How does soap work? ______________________________

  ____

  ____

  ____

Draw a micelle.

Experiment 7: Black Is Black?

Date: ________

Objective

____

____

Hypothesis

____

____

Materials

ball point ink pens of various colors, including black
rubbing alcohol
coffee filters (white)
several small jars
cardboard shoe box (or similar size box)
tape
measuring cup
scissors
ruler

Experiment

1. Pour 60 milliliters (1/4 cup) of alcohol into each of several small jars.
2. Remove the thin plastic tube from the inside of each ball point pen..
3. Pull off the top or cut the end off the plastic tube.
4. Take the ink tube and swirl the open end of it in the alcohol. Make sure that some of the ink gets dissolved, but don't let the alcohol get too dark in color.
5. Cut the coffee filter paper into thin strips 6-12 millimeters (1/4 to 1/2 inch) wide and 13-15.5 centimeters (5 to 6 inches) long.
6. Place the ends of the strips in the dissolved ink in the jars, and allow the alcohol to migrate upwards. It is best if you can suspend the strips in the alcohol without letting them touch the sides. To do this, tape the strips to the inside of a cardboard box, and suspend them in the glass jars. It is OK for the strips to touch the sides of the glass jars, but the alcohol won't migrate past this point.
7. The colors in the ink will migrate up the absorbent strips. Let the strips sit in the alcohol overnight.

Results

Tape the strips of paper below. Write down each original ink color, and record the different colors that each is made of.

Orange Ink made of yellow and red						

Unknown Ink

Repeat the previous steps using a sample of an unknown ink. Try to determine the colors that make up the unknown ink sample by comparing your results with those of the previous samples.

Unknown

Color of the unknown ink:

____

Conclusions

Review

Define the following terms:

sieve ____

____

filter ____

____

filtration ____

____

pores ____

____

solid state ____

____

gaseous state ____

____

liquid state ____

____

chromatography ____

____

separation ____

____

Experiment 8: Show Me the Starch! Date: ________

Objective

____

____

Hypothesis

____

____

Materials

tincture of iodine [Iodine is VERY poisonous—DO NOT EAT any food items with iodine on them.]
a variety of raw foods, including:
- pasta
- bread
- celery
- potato
- banana and other fruits

liquid laundry starch (or a borax and corn starch mixture)
absorbent white paper
eye dropper
cookie sheet
marking pen

Experiment

1. Take several food items and place them on a cookie sheet.
2. Using the eye dropper, put a small amount of liquid starch on a piece of absorbent paper, and label it **Control**. Let it dry.
3. Add a drop of iodine to the starch on the control paper. Record the color in the following chart.
4. Add iodine to each of the food items and record the color for each.
5. Compare the color of the **Control** to the color of each food item.
6. Note those food items that changed color and those that did not.

Results

Food Item	Color
Control:	

Conclusions

Review

Define the following terms:

nutrients ______________________________

____

carbohydrate ______________________________

____

monosaccharide ______________________________

____

disaccharide ______________________________

____

polysaccharide ______________________________

____

starch ______________________________

____

cellulose ______________________________

____

amylose ______________________________

____

amylopectin ______________________________

____

Experiment 9: Gooey Glue

Date: ________

Objective

____

____

Hypothesis

____

____

Materials

liquid laundry starch (or a mixture of borax, corn starch, and water)
Elmer's white glue
Elmer's blue glue (or another glue different from white glue)
water
2 small jars
marking pen
Popsicle sticks for stirring
measuring cup

Experiment

Part I

❶ Open the bottle of Elmer's white glue. Put a small amount on your fingertips. Note the color and consistency (sticky, dry, hard, soft) of the glue. Record your observations.

❷ Now look carefully at the liquid starch. Pour a small amount on your fingers or in a jar. Note the color and consistency of the starch. Record your observations.

❸ Take one of the jars, and put 60 ml (1/4 cup) of water into it.

❹ Note the level of water in the jar, and draw a small line with a marker at the water level.

❺ Add another 60 ml (1/4 cup) of water, and mark the water level with a marker.

❻ Pour the water out.

7. Fill the jar to the first mark with Elmer's glue.
8. Fill the jar to the second mark with liquid starch.
9. Mix the glue and the starch with a Popsicle stick. Record any changes in consistency and color.
10. Take the mixture out of the jar, and knead it with your fingers. Observe the consistency and color, and record your results.

Results

- Observations for Elmer's white glue:

____

____

____

- Observations for liquid starch:

____

____

____

- Observations for the mixture of Elmer's white glue and equal amount of liquid starch:

____

____

____

____

____

Part II

1. Take another jar, and follow Steps 3-6 in Part I of this experiment. This time, fill the jar to the first mark with the Elmer's blue glue or another glue that is different from the white glue.
2. Add liquid starch to the second level.
3. Mix.
4. Record your observations.

Results

- Observations for mixture of blue glue and liquid starch:

____

____

____

____

____

____

____

____

____

____

____

____

____

____

____

Conclusions

Review

Define the following terms:

meros ____

____

polymer ____

____

monomer ____

____

polyethylene ____

____

vulcanization ____

____

Experiment 10: Amylase Action

Date: ________

Objective

____

____

Hypothesis

____

____

Materials

tincture of iodine [VERY POISONOUS—DO NOT EAT any food items that have iodine on them]
bread
timer
wax paper
marking pen
cup

Experiment

1. Break the bread into several small pieces.
2. Chew one piece for 30 seconds (use the timer), chew another piece for 1 minute, and a third piece for as long as possible (several minutes).
3. Each time, after chewing the bread, spit it onto a piece of wax paper. Using the marking pen, label the wax paper with the length of time the bread has been chewed.
4. Take three small pieces of unchewed bread, and place one next to each of the chewed pieces.
5. Add a drop of iodine to each piece of bread—chewed and unchewed.
6. Record your observations on the following chart.
7. Take two more pieces of bread. Collect as much saliva from your mouth as you can (spit into a cup several times). Soak both pieces of bread in the saliva. Place one piece in the refrigerator, and leave the other piece at room temperature. Let them soak for 30 minutes.
8. After 30 minutes add a drop of iodine to each. Record your results.

Results

Chewed Bread				Bread + Saliva 30 minutes	
30 seconds	1 minute	Several minutes	Unchewed Bread	Refrigerated	Not Refrigerated

Conclusions

Review

Define the following terms:

protein ______________________________

____

amino acid ______________________________

____

peptide bond ______________________________

____

kinesin ______________________________

____

DNA ______________________________

____

nucleotide ______________________________

____

double helix ______________________________

____

DNA polymerase ______________________________

____

Draw a picture of kinesin.

What are the four bases that make up DNA?

______________________________ ______________________________

______________________________ ______________________________

What are the symbols for the four bases that make up DNA?

____